D1456176

ON TREMULATION

BY

EMANUEL SWEDENBORG
(1688 – 1772)

TRANSLATED FROM THE PHOTO-LITHOGRAPHED COPY
OF THE SWEDISH MS.

BY

C. TH. ODHNER

BOSTON
MASSACHUSETTS NEW–CHURCH UNION
16 ARLINGTON STREET
1899

Swedenborg, Emanuel

Reprinted in 1976   400 copies
Swedenborg Scientific Association
Bryn Athyn, Pennsylvania

TO

THE SWEDENBORG SCIENTIFIC ASSOCIATION

THIS VERSION

OF SWEDENBORG'S EARLIEST PHYSIOLOGICAL WORK

IS RESPECTFULLY DEDICATED.

Bewis at wårt lefwande wäsende består merendehls af små darringar, thet är, Tremulationer.

(Arguments showing that our vital force consists mostly of little vibrations, that is, Tremulations.)

[TITLE OF THE CLEAN COPY HANDED IN BY THE AUTHOR TO THE BOARD OF HEALTH, OF STOCKHOLM, IN 1719.]

Anatomi af wår aldrafinaste natur, wisande at wårt rörande och lefwande wäsende består af Contremiscentier.

(Anatomy of our most subtle nature, showing that our moving and vital force consists of Contremiscences.)

PREFACE.

EMANUEL SWEDENBORG'S treatise, "On Tremulation," which now for the first time appears in the English tongue, was originally written toward the close of the year 1719, as may appear from the following statement in a letter by the author, dated Nov. 3, 1719, and addressed to his brother-in-law, Dr. Eric Benzelius, then librarian of the University of Upsala : —

I have also written a little anatomy of our vital forces, which, I maintain, consist of tremulations; for this purpose I have made myself thoroughly acquainted with the anatomy of the nerves and the membranes, and I have proved the harmony which exists between that and the interesting geometry of tremulations; together with many other ideas, where I have found that I agree with those of Baglivius. [Giorgio Baglivi, a disciple of Malpighi, and professor at Rome.] The day before yesterday I handed them in to the Royal Medical College. (See R. L. Tafel's "Documents Concerning Swedenborg," Vol. I., p. 310.)

From the contemporary entries in the Proceedings of the *Sundhets Collegium*, or Board of Health, in Stockholm, it appears that this work of Swedenborg's was duly received and reported, the Board resolving that the treatise should be read in turn by all the members, who afterwards were to pronounce an opinion respecting it. While thus circulating, it seems that the manuscript disappeared, as there is no further reference to it in the Proceedings of the Board, and as it has not been preserved in the library of the Royal College of Medicine at Stockholm. Swedenborg himself retained only the first rough draft, which has also disappeared, but from it he made a second copy of chapters I.— VI., and XIII., which fortunately has been preserved. This, therefore, is all that remains of the original work, "On Tremulation," and it is from this copy that we have prepared the present translation.

A few quotations from Swedenborg's correspondence with Benzelius will give the history of this second copy, and also illustrate the nature of the work itself : —

[Stockholm, middle of January, 1720.] By the last post I began sending over to you my latest literary efforts. I should be very glad if this, as well as what is to follow, meet with your approbation. It is certainly true that Baglivius first started the theory; and that Descartes treated upon it, and afterwards Borellus; but no one has yet furnished any proofs, or treated the whole subject fully; wherefore I claim my proofs as new and as my own, although the subject, or the theory itself, I am willing to leave to others. Still I must say that a great part of what I discovered myself I afterwards found I had done in conjunction with Baglivius, which has rather pleased my fancy; as, for instance, what I have to say about the function of the meninges. The whole will cover a large space; I think it will occupy seven or eight weeks, even if I send you portions twice a week. The physicians here in town will take the subject into consideration, and all express themselves favorably. (Documents, I., 317.)

[Stockholm, February 24, 1720.] I break off my article now, and send chapter XIII., lest there might arise a squabble [among the professors in Upsala] as to the proper meaning. It would be very desirable if, in the objections that may be raised, respect were had to such things as would contribute to set this matter in its proper light for me; I mean that such objections should be raised, by which I might in a certain measure see whether I am on the right or on a wrong track; but merely to imagine many things about the animal spirits, and to admit only such things as have reference to their chemistry and function, and none that concern their geometry, seems too weak a defence. For I lay it down as a principle, that the tremulation begins in the fluid which is contained in the membranes; in order that this tremulation may spread, the membranes require to be in a state of tension with the hard substances as well as with the blood vessels; for in such a case all the lymphatic vessels, or the vessels of the nervous fluid, lie upon the membranes in their proper order, and exert a pressure upon their contiguous parts in an instant, just like any other fluid, and they thus communicate a trembling motion to the membranes, and also to their bones; so that almost the whole body is brought into a state of subtle co-tremulation, which causes sensation. I presume that Messieurs, the Academicians, are so reasonable, as to set aside childish prejudices, and oppose reasons to reasons, so as to see on which side is the greatest weight. (*Ibid.*, p. 318.)

[February 29, 1720.] I send you now the continuation of the preceding part. I wish much that it may gain the approval of the learned who are concerned in the subject; but as I am doubtful of this, I will allow some interval to elapse, that I may learn meanwhile what objections may be raised to it: for if any one entertains an opposite opinion, the best arguments may be thrown away; in preconceived opinions every one is almost totally blind: still I will with all my heart leave to your good pleasure, and to the service of the public, anything that may be demanded. Care must be taken not to draw down upon oneself the anathemas of the learned, on account of new discoveries, or some hitherto un-tried argumentations. In the next chapter there seems to me to be contained better and more evident proofs, which are taken from the senses and our sensations. I have some other parts besides, which are not yet worked out, and which treat of the mechanism of our passions and the movements of our senses, so far as they may be deduced from the structure of the nerves and the membranes. To this there will be added some unknown properties possessed by the least ramifications of the arteries and veins, for the purpose of continuing motion; but inasmuch as this requires to be established by several courses of thought, and by anatomical investigations, I reserve it for some future opportunity. . . . The whole of what has been sent over to you has been written off from the first draught; should any mistakes have crept in with regard to the orthography, you will please attribute it to the fact, that a proper copy does not yet exist. (*Ibid.*, p. 319.)

[Brunsbo, April 12, 1720.] Since my departure from Stockholm, I have not had time to send you the continuation of my Anatomy; nor can I send it to you from here, because I have not my first draught with me, and my head does not well recall things from memory; with the first opportunity I will again communicate something to you. (*Ibid.*, p. 324.)

[Brunsbo, May 2, 1720.] It would be my greatest delight if I could continue my Anatomy from here. The first draught was left at Starbo, and without it it would make my head ache, to endeavor to hunt up the various threads which are already deeply *obducta alius generis cogitationibus* [that is, covered up by thoughts of a different kind]. Still it shall be done, as soon as an opportunity offers. (*Ibid.*, p. 325.)

This is the last reference made by Swedenborg to his little work, "On Tremulation." The desired opportunity did not offer itself, and Benzelius, consequently, never received any further instalment of the work. The copy of chapters I.—VI., and XIII., was subsequently carried to the city of Linköping, when Benzelius, in 1731, was appointed Bishop over that diocese, and there it remains until the present day among his other papers, which are preserved in the library of the cathedral. Dr. R. L. Tafel, in 1869, procured a photo-lithographic copy of the manuscript, which constitutes pages 132–180 of the first volume of Swedenborg's photo-lithographed manuscripts.

As indicated by the author, this copy was transcribed from the first rough draft, which will account for certain unpolished sentences and other crudities of diction, many of which will be apparent also to the English reader. The original language is very peculiar, indeed, both as to orthography, syntax, and vocabulary. The Swedish of the early part of the eighteenth century was as different from modern Swedish, as was the language of Tyndale or Coverdale from modern English. Swedenborg himself was, in fact, one of the first who ventured to employ Swedish in a scientific treatise, and he was therefore forced to coin many new and strange expressions, and to borrow largely from other tongues, such as the Latin, French, German, and even English, with which the original of this work is plentifully besprinkled. The author himself soon recognized the insufficiency of the Swedish, as then existing, as the vehicle of scientific thought, and therefore, in all his subsequent works, he fell back upon the all-dominant Latin.

Leaving to the reader to judge of the intrinsic value of the present treatise, we will merely point out its historical importance as a contribution toward a correct understanding of the growth of Swedenborg's mind, and of the beginnings of those great principles of natural truth which received a more perfect development in his later scientific and philosophical works. It is to be noted that this treatise was written when the author was but thirty-one years of age, and that it is the first of all his anatomical or rather physiological works. It may be regarded as distinctly marking the close of the first period of Swedenborg's career as an author and scientist. During this period, which commenced in the year 1709, he had written no less than twenty different treatises, nearly all in the Swedish language.

Some of these were published by himself, and all have
been preserved in one form or other, but none of them
has as yet appeared in English. All may not be of
supreme value, regarded in themselves, but they are
nevertheless indispensable to a thorough comprehension of Swedenborg's preparation for that unique and
stupendous mission which awaited him. Beginning
his literary career as an annotator of the classics, he
next appears as a Latin poet of no mean ability. For-
saking Polyhymnia for sterner muses, he now delves
into mineralogy, geology, astronomy, mathematics, and
physics, writing numerous interesting and suggestive
little works on all these subjects, while at the same
time publishing his *Dædalus Hyperboreus* or Journal of
Mathematics, Mechanics, and other physical sciences.
In the sixth and last number of this journal, which was
written in the beginning of 1717, but not printed until
October, 1718, we find an article on the subject of
Tremulation, which we have added as an introduction
to the present treatise, being the conception and fore-
runner of this more extended work, which may be
looked upon as the last work of Swedenborg's youth.

He now appears to have begun his studies and labors
over again, as it were, in a more thorough and system-
atic manner, and with more mature results. Leaving
physiology for a time, he returns to metallurgical, geo-
logical, and astronomical subjects ; he writes his " Lesser Principia " in 1720, publishes his " Chemistry " in
1721, his " Miscellaneous Observations on Minerals,
Fire," etc., in 1722, and the " Principia," the " Regnum
Subterraneum," and the " Outlines on the Infinite," in
1734. Having thus a second time run through the

cycle of these more ultimate sciences, Swedenborg, in 1735, resumes his study of the human body, which he fitly terms " the temple of all the sciences." The great work, "On the Brain," the " Economy of the Animal Kingdom," the " Rational Psychology," the " Organs of Generation," " The Animal Kingdom," and others now follow one another in rapid succession, but through all of these magnificent works of philosophic science there vibrates the key-note which many years before was struck in the work, "On Tremulation." Nay, even in Swedenborg's latest theological writings there will be found many traces of the principles and arguments first presented in this little treatise.

C. TH. ODHNER.

HUNTINGDON VALLEY, PA.
February 15, 1899.

[From *Dædalus Hyperboreus*, No. VI., October, 1718.]

ARGUMENTS SHOWING THAT OUR VITAL FORCE CONSISTS MOSTLY OF LITTLE VIBRATIONS, THAT IS, TREMULATIONS.

BY ASSESSOR EMAN. SWEDBERG.

BEFORE what is unusual and unknown can be made credible, it is necessary to establish some fixed and indubitable rules, according to which the theory may be proved.

THE FIRST RULE OF TREMULATION.

Anything of a firm and hard nature, such as wood, stone, rock, metal, etc., is subjected to great tremulations even by a slight touch.

This is evident from buildings and cities: houses and streets are known to tremble and reverberate from a wagon passing by; a whole rock trembles at the knock of a hammer; a bell vibrates and even produces sound from the touch of a small needle; a person at one end of a long pole or mast may know what another person is writing or drawing on the other end; if poles were joined one with the other to the length of a mile, or in the tube of a draught-engine, a blow would be noticed from one end to the other; nay, even if only one end should touch a stringed instrument, the vibration would at once be communicated to the other end; a cannon-shot, a mountain-slide, a subterranean cave-in may be heard twenty to thirty miles round about, causing houses and cities to tremble and shake. From this it may be concluded that a small cause is able to effect a great vibration.

The Second Rule.

An expanded membrane is the best medium of tremulation.

It is known that a membranous string is the best medium of sound, that is, of tremulation. By a membrane is meant anything most external or the surface of a solid substance, which surface receives the tremulation before it is received in the body itself, which consists of continuous membranes and surfaces; and this in the same ratio as that of the square to the cube.

The Third Rule.

Next to membranes, the best media of tremulation are such bodies as are hard and elastic; softer bodies are less suitable.

The most brittle and hard metals, such as iron and steel, or copper and tin together, are the ones that give a ringing sound; the more plastic metals, such as gold and lead, give less sound; softer substances, such as sand, clay, or feathers, give no sound whatever.

The Fourth Rule.

The tremulation of a string will cause a sympathetic vibration in another string; a membrane similarly affects another membrane; that is, if both are tuned in the same key.

If the string of a lute is touched, it will cause a vibration in the other strings which are tuned in the same key. An outside sound will often cause a vibration in a whole musical instrument, as also a whole gallery will vibrate from the sound in the pipe of an organ, that is, if they are in the same key or tune. A glass may break from its own sound.

THE FIFTH RULE.

Tremulations in the air make rings and circles, and are heard on all sides round the center of the motion; that is, if the whole mass is not being moved.

If a stone is thrown into the water, it will make rings round about. So also, in the air, a cry or sound is heard on all sides round about.

THE SIXTH RULE.

The heavier the atmosphere, the slower is the tremulation, but the lighter the air, the swifter is the motion.

The tremulatory circle moves slowly in the water; in the air it moves more quickly; in the finer air, which is called the ether, it is still swifter; in the solar substance it moves from the sun to us in an instant; in the very finest atmosphere there is probably no time which can correspond to the undulation.

THE SEVENTH RULE.

One tremulation does not interfere with another, simultaneous one.

This may be best tried in water, where ten or twenty circles may oscillate the one within the other, without

interfering with one another, but each one proceeds on
its way without obstruction.   Similarly in the air: the
sound of one string of an instrument does not interfere with the sound of another string, nor one word
with another.   The reason of this will be shown sep-
arately.

THE EIGHTH RULE.

*In all tremulations the angle of reflection is equal
to the angle of incidence.*

The oscillating circles in the water are reflected according to the angles of incidence.   In a round vessel
full of water they return to the center ; in an oblong
channel they go forward and backward ; a rope, hanging in a mining-shaft, moves itself up and down in serpentine coils ; the same takes place in a musical chord ;
an echo propels the sound forward and backward ; thus
also does the substance of the sun move the particles
of our sight.

THE NINTH RULE.

In tremulations there are millions of variations.

How many different sounds are not produced by a
well-tuned piano ? how many are not still lacking within
an octave ?   One sound is different from the other, is
more flowing, broader, duller, or harsher.   The sound
and pronunciation of men differ like their faces.
Every vowel has its own separate sound.   If these
variations are duplicated, it will be seen that there are
millions of different kinds of tremulations.

THAT MUCH OF OUR VITAL FORCE CONSISTS IN
TREMULATIONS. *See pages 6 and 7 etc.*

From the above rules it can be shown that our mobile life, or our nature, consists in little vibrations, that is, tremulations. From the first rule it may be seen that a most minute particle is able to communicate its motion to all other things in the whole body, is able to bring a certain membrane and sinew, the blood, the life, and the spirit into the same motion with itself, and thereby all contiguous membranes, fibres, and nerves. *Speech* is nothing but tremulation, like the sound in a string. *Hearing* is only a concentrated collection of such tremulations, flowing in through turbinated membranes, and propelling themselves over hammers and anvils up to the *dura* and *pia mater*, which are similarly vibrating; so that all fibres, nerves, animal spirits, and the blood, thus touched, will come into a motion according to the preceding rules. If the same tremulation is caught by the membrane by means of the sympathetic vibration of the teeth or the bones of the head, the sensation of hearing may be effected independently of the mechanism of the ear. *Smelling* and *taste* are similarly produced by contact with various kinds of particles — round, angular, or sharp — by which the fibres and nerves are pulled or drawn, carrying the tremulations to the *dura* and *pia mater*. Similarly with the *Sight*, which is the most delicate of our senses; the least of color or light strikes against the minute fibres, and the distended optic nerves communicate it to the coats of the brain, effecting sensation and tremulation round about. It is the same with the motions of the temper, which are derived from a stinging or biting of the bile in the internal organ; similarly in the case of all external feelings or sensations, because all things are so connected by threads and sinews, that what is touched in one place is felt in another, and especially in the membrane of the brain, for all the threads and nerves terminate there, and into it is collected whatever belongs to the whole body; if, therefore, the tremulation is first felt in that membrane, it will at once find space and matter by means of which to communicate itself over the whole body. If now this membrane becomes slack, or is deprived of its heat, its blood, or its animal spirits, then the whole man becomes dull, heavy, and dead. Further, when a relaxation takes place in the nerves of the five senses, after having been in a state of tension during the entire day, then sleep sets in; and yet, during the sleep, we have something similar to sight and hearing, etc., which things go to prove that all the external senses are still kept in internal tremulations. It also frequently happens that a person falls into the thought of another person, that he perceives what another is doing and thinking, that is, that his membrane trembles from the tremulation of the other person's cerebral membranes, just as one string is affected by another, if they are tuned in the same key. It may not be presumptuous to conclude that the thoughts of the unreasoning animals are tremulations, proceeding from the internal and external sensations of the body and its senses; so that experience has taught them what is meant by one kind of tremulation and what by another, just as we recognize the words and their meanings by the different kind of tremulation in each sound.

Note
(1)

→That no part of the body can be touched without communicating the touch to the *dura* and *pia mater*, that is, the membrane of the brain, and that nothing can be touched in this membrane, without communicating it to the whole body — may be seen from all those threads which are joined to the sinews, outwardly; as also from those twenty to thirty nerves which terminate in the cerebral membrane, and which all are clothed by it. If said membrane is injured, a person is at once deprived of a sensation; he fails, swoons, loses his thought and his reason. If the fibres or nerves become slack, a person is similarly deprived of sensation, as takes place in colds, congestion of the blood, or in sleep. If the animal spirits are overflowing, as in intoxication or in anger, then the membrane becomes too much heated and distended, so that it makes a manifold and wild tremulation instead of the proper and usual one. From all this it follows, that, by means of so many contacts or impressions, there is in us a continually moving, tremulatory, and living force, in the leasts as in the greatest, according to the preceding rules.

Experiences in man through characteristics of matter or substances in his anatomy are constantly at play, laying a base for feeling values, & thinking logic and even being recorded in both the conscious and unconscious sensative areas or circles of the brain ready to be utilized by the directive mind of each human.

An important aspect of what might be called the spherical mind. "As above so below" from larger greater spheres to and through smaller ones down to the least cells of many kinds (animal, plant, embryo, electrical, chemical).

15 May 1981

Much needs to be researched to make clear the interrelation of a multitude of things, if not all things. How magnificent of the Oral

embryological

ARGUMENTS SHOWING THAT OUR VITAL FORCE CONSISTS MOSTLY OF LITTLE VIBRATIONS, THAT IS, TREMULATIONS.

CHAPTER I.

§ 1. IF common sense be consulted and allowed to guide us as we inquire further and further into the real cause of life — as to what it is that really makes us living and wherein living force most generally consists — we must finally come to the conclusion that this cause is motion. For is it not according to common sense that everything that lives also moves, that is, that the living or the being is inseparable from the moving?

Life consists both of certain internal senses and of a number of external ones. No one can deny that the *external senses* owe their existence to motion, for there must be something in the atmosphere which flows in with certain little impulses and circlings, moving about the finest fibres and most minute termini, which by means of tremulation or little vibrations carry forward the motion to a certain sensation, and which thus by a motion distributed over an entire system contribute or effect together a symbolum of life. Thus also with the *internal senses :* what thought is there, or what living recollection, in which motion does not effect as well the first impression as the last? In a word, if common sense is followed, we will inevitably find that *Rest* can never have any part in that which is called *Life;* for

rest and life are two contrary things, just as a dead
state and a living state.   Experience testifies to this:
as soon as the motion is obstructed by any obstacle, it
is seen that life is at once deprived of a certain spark
of its proper nature ; but as soon as something more
moving is added, it is seen that the liveliness is increased.   This may best be seen from insects and
other small animals, with whom life resides as it were
in the least little drops of fluid : if the sun-ray strikes
them, or touches their fibril, membrane, or little vein,
then their whole life quickens at once and is as it were
kindled, and the senses begin to live ; but this motion,
or this life, ceases as soon as the cold season arrives,
for cold is the very opposite to motion ; life with them
is therefore mostly a motion.   With the greater bodies
the cold cannot effect so much, for it has less power
to penetrate a great solid, as may be seen from geometry, so that we are not deprived of motion during
winter, but still the principle is the same.   Cold
weather often causes what is most external to become
dormant or extinguished ; a person is chilled, or is deprived of a certain sensation, so that a loss of spirit,
or a dead state is always the result when the motion is
stopped by the cold.   In a word, life consists in the
motion, but death in the *rest* of the particles.

Now in regard to the finer motions which cause that
we live, that we have the use of our senses and our
thoughts, and that we possess the complete harmony
or communication of all these things as one, it should
be remembered that these motions are of a more subtle
essence than those which have been examined by the
learned.   The geometry of these motions is closed to

us and to our coarser senses, so that we can hardly be
said to have come further than to the first step of the
knowledge concerning them, many thousand steps still
remaining before we will be able to ascend to any perfect knowledge.   For all that which makes *the being* of
a sense, is more subtle than the sense itself and whatever is effected by that sense, so that it seems that
only a finer sense is able to form a judgment concerning a grosser one, but the latter cannot form any judgment regarding itself.   The ear, for instance, cannot
possibly know or feel what it is that is vibrating in its
organ, or how one thing is moving against another,
unless a more subtle organ reveals it.   The thought,
which mostly is kept in attention to the feelings of the
external senses and in their collective center, is not of
itself aware of that which constitutes its own motion
and life.   In any case the conclusion must be this, that
those motions in which life resides are the most subtle
of all motions, of a nature such as cannot be seen or
comprehended by any comparison with the grosser
forms of motion.

Note well

§ 2.   Tremulation is the most subtle form of motion
that exists in nature, and it possesses wonderful and
distinctive properties, differing from all other motions.
Although what is tremulatory presents itself each moment before us, playing round about each of the senses,
yet is our mechanism and our reason still so little cul-
tivated, that we have no proper knowledge of tremulation and its most subtle nature, as to wherein it consists, and wherein it differs from other motions.

If tremulation is closely examined it will be found
that it most closely resembles an axillary motion, as to

Note
well

its subtlety, or a motion within the least of space, that is, such a motion as takes place at the centre alone; and that it has hardly anything in common with local motion, which takes place from one place to another. Tremulation, consequently, is not subjected to the laws that govern local motion. In a hard substance tremulation seems to be nothing but something swiftly moving up and down, an effort to recover the balance, like a ball thrown against the floor which makes smaller and smaller reboundings, until finally it returns out of the balance of motion into an equilibrium which is in a state of rest. It is the same with the most minute particles which possess hardly any weight and which of themselves move neither up nor down; if touched by the least motion these will leap and bound and tremble, until, after a period of tremulation, they finally return to their rest. Experience shows also that the lighter and more subtle the particles are, the swifter is the communication of the tremulation. Water trembles so slowly that the tremulation can be followed by our observation. Air moves more swiftly, and ether more swiftly still. Fire, or its radii, moves so swiftly that the tremulation is almost instantly communicated to us from the sun itself. It may be seen from this that the whole nature of tremulation consists in the effort of a thing to recover the balance which it was about to lose.

Tremulatory motion has in itself nothing in common with local motion, for it will be found that the latter requires its own fixed times, corresponding to its distances, adding to this, each moment, an increase in a certain measured ratio; it possesses a certain quality in a heavy substance and another in a light one, it is different in relation to a greater surface from what it

Doubly subtle is this action

is in relation to a smaller one; while on the other hand a tremulatory motion can exist in the same thing that is simultaneously subjected to local motion. A thing can be carried from one place to another, while at the same time it is trembling continuously without the least hindrance from the local motion. A bomb flying through the air, may in its course be subjected to tremulation; nay, one tremulatory motion may be within another one; a greater motion may exist together with a lesser one; within the latter there may be a still smaller one, and finally one most minute. Over an undulation, such as is seen in the water, there may be moving a smaller undulation or oscillation, over the latter a tremulation, over this a contremiscence, and so, finally, a most subtle one, which almost might be called a sensation or a *vivum*. A human body on board a ship may undulate up and down with the waves, while at the same time the brain possesses its own undulation; over this again there may be the vibration of a tremor, over this a tremulation, and over this a least trembling, such as produces the sense of hearing. The one motion may therefore be within and above the other, each without interfering with the other. The more a body is stretched out or expanded or in a state of quiescence, and the more fixed and heavy it is, the more does the tremulatory motion seem able to exhibit its proper nature; a whole mountain, entire houses and cities, with bells and belfries, tremble and shake from comparatively small causes, whence it may be seen that tremulatory motion has no consideration for what is heavy and great, but the greater and heavier a thing is, the more freedom does this motion possess to penetrate the whole and to bring everything into a sympathetic trembling, as shall be shown better in what follows.

§ 3. As now living force is motion, and as life consists of little motions, and as the most subtle motions in nature are contremiscences, it follows that whatever lives in us consists of contremiscences, that is, most subtle motions; it is therefore our opinion that whatever lives in us is a tremulation in our finest nerves, in the most delicate membranes, in the very bones and in the entire systems of nerves and bones. A sensation of hearing, for instance, is first produced by a motion in the air and then in the membranes of the ear; this is then communicated from one membrane to another, from one nerve to another, from one bone to another; and as all the membranes are connected one with the other, and as the membranes and the nerves join and make a common system, it follows that the least tremulation in a nerve or a membrane is able to distribute itself over the whole connected system of the body. The bones, also, are joined one with the other, and each is surrounded with its own membranes, so that as soon as a tremulation enters into a bone, it flows at once over the whole osseous system, as shall be proved in what follows. Our theory is therefore, that every part of what is living in the body lives by means of little tremulatory motions which flow into the nerves and the membranes and set the whole system into sympathetic tremulation; and that as soon as a contremiscence is distributed over a whole body, it may be termed a sense or a sensation, and that if all the contremiscences of the senses are taken conjointly, they possess the name of nature, or of life. This, then, is what is to be demonstrated.

CHAPTER II.

§ 1. In order to form a correct judgment of the more subtle motions in our organism, and of the more invisible contremiscences, and in order to show where life really resides, let us consider the testimony of the greater undulations in our body. The greater motions must be the origins of the lesser ones, and the beginnings of tremulations can be found only in undulations. *The lungs*, in the first place, are the fountain of a multitude of motions ; here the external air is first received ; the inflated organ makes greater or smaller expansions and communicates its motions to everything in its connection, and these motions are undulations, which are nothing but a grosser degree of tremulation. *The heart*, also, has its own motion : wringing and twisting itself in and out, it propels the blood through the arteries into the veins, and thus, by its pressure, it effects the circulation throughout all those bloodvessels and channels of which the body chiefly consists. While it is true that this wringing motion of the heart cannot properly be called an undulation, yet, so far as the circulation is a reflected movement, going and rebounding to and fro, even though by a circular course, it may still be named an undulatory or vibratory motion, like the motion of a horizontal pendulum, though here somewhat by a spiral. *The brain* similarly possesses a reciprocal or undulatory motion, which accommodates itself to the wringing motions of the heart, or it may be that the latter moves in obedience to the undulation of the brain ; moreover, it has been discovered in our own age that the medullas, both the oblongata and the spinalis, vibrate and respire and rise and fall as if in fermentation. If now hypotheses be allowed to mingle with anatomical experience, we may easily suppose that these undulations propel a fluid into the nerves (just as takes place with the blood in its more open and hollow vessels), and thus to the extremities or membranes ; these membranes reciprocally returning the fluid through the nerves back to the meninges of the brain, over which expanses the tremulation flows in the first instance. The brain, therefore, is a fountain, whence flows a fluid through the nerves to the membranes, keeping the latter expanded and in proper condition. This, then, goes to show that nature is everywhere endeavoring to communicate life by means of a circulation, and especially by means of undulations, that is, by greater or lesser motions of tremulation.

As was said, nature, and consequently all that is living in us, is in the effort to proceed by help of tremulations. As a proof we may first adduce the testimony of what is visible. Consider *speech*, for instance : words are to be expressed and sound is to be communicated, by distinct articulation, to the hearing and understanding of another person ; — this, then, is effected by little atmospheric vibrations which are formed between the folds of the tongue and by means of the air being strained or filtered among these folds, as also through other turnings and twistings, all of which make the tremulations distinct and articulate. If the sound of *A* is to be expressed, the palate and the whole formation of the mouth know at once how to open the way to let the sound flow forth in a different manner than if a *B* or another sound is to be produced ; these being some of those twenty or more varieties of tremulation which can be formed in our mouth. This is therefore a proof that tremulation produces everything of speech.

Tremulation, moreover, often shows itself throughout the whole nervous system, to such a degree that it is called *tremor, shivering, convulsion ;* these are really nothing but coarse tremulatory motions in the whole nervous system, showing that the nervous body is disposed to permit the tremulation to play freely over its field, and that there is such a conjunctive communication between all the parts that a tremulation is distributed over a whole system as soon as a single nerve is touched. It has been observed that this kind of tremulations shows itself when the membranes become empty of blood, or when something in the nerve becomes torpid, so that the fluid cannot play in proper freedom, or if a membrane or nerve is injured or loses its usual tension and becomes slack. Hence it is evident that a single tremulation may in a moment spread over entire systems, and thus over that whole part or body which is in a state of tension.

Tremulation is exhibited in a somewhat less degree in the periostea, and is often noticed as a delightful contremiscence beneath the pericranium, as it were ; this is occasioned by some pleasantness which plays and titillates in the mind, producing an harmonic motion with such a degree of excitation that it makes itself felt even in the membranes of the external brain ;

the same, also, is often occasioned by a sudden feeling
of astonishment which is mixed with a certain degree
of fear, when one may consciously feel how the tremulation flows over certain membranes like a wave of
cold water, and over the head like the most delicate
undulations beneath the roots of the hair, which then
often feel as though they were rising and standing on
end ; whence there is distributed over the whole body
a tremor or passion often growing into a greater trembling. This sensible contremiscence must necessarily
accompany the motion which is taking place within the
dura mater itself, for the pericranium is known to be
so joined with the interior meninges by little fibrils and
tendons, that whatever takes place in the interior, must
become sensible also in the exterior. Such a sensation, therefore, is often the alternate of a passion which
proceeds by a tremulation, gradually increasing, dis-
tributing itself over the matres in the body, and consequently communicating itself to the nervous system
and to the membranes of the periosteum.

The tremulation, in a still less degree, may be felt
by a person who falls into *a passion*, no matter of what
kind : when the passion has cooled off, there follows a
contremiscence in the whole nervous body ; but if the
mind is to be able to reflect properly upon this subject,
it must not continue in the disturbed state which necessarily prevents the tremulation from becoming sensible, but it must fall into a state of tranquil thought,
and then in each finger, and in each limb, there will
be felt an internal tremulation over the whole body, as
will be described more particularly toward the end of
this treatise. All this, therefore, is a sensible proof

that the tremulation endeavors to ultimate itself in the whole nervous system, or rather, in the whole body, and it shows that nature must express itself by means of tremulation, as in the greatest things, so in the leasts.

§ 2. Touching the geometry of tremulation, it is to be observed that there are various degrees of this motion, greater and smaller, just as in local motion there are degrees as to swiftness and slowness. Local motions can be so slow that the sense of sight cannot observe any changes as to distances, except through a long period of time; such motions, for instance, as those of the hands on a watch, which point out the hours, and measure days and months. It is the same with those local motions which the stars and planets are making before our eyes by their orbits in the universe; unless we calculate such motions by the help of time, our senses might imagine that these bodies had no motion whatever. Local motion may, on the other hand, be so subtle and swift that it vanishes from our sensation, as in the case of a bullet which travels through the air and traverses our sight so swiftly that we can make no observation of its course. And yet both the slow and the swift are local motions.

Our tremulatory motions possess similar degrees of swiftness. The most sensible and visible, which is the first degree, is *undulation*. This, as has been said before, is exhibited in the greater motions of our body, which are the motive springs of the minor motions which may be designated tremulations. All this may be compared to a great wheel governing a thousand minor wheels by which it effects the motion of the whole machinery.

The same greatest degree of tremulation, which is named undulation when its swinging motion may be distinctly seen, can be observed in many other things: a ship's mast, resting on its keelson and with the other end in the air, will undulate from the least cause, seeking to regain its rest or balance, which was about to be lost by the weight or overbalancing of the top; if an elastic ball is thrown against the floor, it will rebound and make undulatory reflections up and down, these reflections gradually growing smaller and smaller, until finally, by increasing efforts, the ball returns to its own rest and balance; a pendulum, which is left to move freely in the wind, will also make a vibration or a horizontal undulation, and this motion is altogether of the same character as a tremulation, as will be described in what follows, for the better understanding of the nature of tremulation. If an element, such as water, comes into a state of undulation, it drives out waves and rings on all sides round about, and makes circular oscillations further and further from the center, presenting a visible tremulation, similar to the motion in the air which produces sound. The latter, as has been said, flows up and down, making smaller and smaller waves, until an equilibrium has been restored and an even quietude reestablished. From all this it may be seen, that the tremulatory motion, no less than local motion, possesses a greatest degree of slowness, and that it is this degree which is termed undulation.

The second degree of this motion commences at the boundary in which the undulation terminates. As soon as an undulation begins to become audible — however coarse or dull be the sound — then begins that swifter degree of undulatory motion which is properly termed *tremulation,* and which embraces all that sphere of vibrations that is produced by sounds and chords. For it is the greater or lesser swiftness of the motion that causes the sound in the air to be heard, or makes the tremulation to reverberate in the air, communicating it to the tympanum and the other membranes; a less degree of swiftness produces the grosser and duller sounds, while a greater degree makes the finer sounds, until the swiftness vanishes in such a subtlety that the tremulatory motion again escapes the observation of the organ of hearing, just as the local motion finally escapes the organ of sight, as was said above. Within this degree of tremulation we must, therefore, include all sounds, from the deepest to the highest; by certain experiments it may be observed that at least one hundred and fifty vibrations in a second make the highest *c* in a piano, while thirty or forty vibrations produce the lowest *c*; certain tremulations, of the most slow or undulatory character, may even be noticed by the eye, like a mist around the string, but the finest tremulations of the string escape the eye and finally also the ear.

The third degree of tremulation begins, therefore, where the vocal or harmonic vibration ceases or vanishes from sight and hearing, and this degree should be termed *contremiscence* or sensation. Commencing when the vibration becomes more rapid than two hundred in the second, it may increase in swiftness until it reaches one or two thousand within the same time, when it is no longer to be followed by the sight or hearing; the other senses, therefore, such as taste,

smell, and touch, must then assist in order to compre-
hend the sphere of this degree of tremulation and to
apply it to our motive life-force, as shall be further developed in what follows.

§ 3. Our conclusion is, therefore, that our whole living and moving nature endeavors to express itself by
means of tremulations.   The greater of these motions
keep all the grosser parts of the body in an even movement, as also in a state of tension and expansion for
the reception of the minor motions ; the latter, again,
are in themselves new motive forces and the sources
of the motions of the finer parts of the body, and of
the most delicate degree of tremulations.   This may
be illustrated by a clock-work, in which a hundred different little wheels are set in motions by the motion of
one greater wheel, or even by a single vibration of the
pendulum.   It is the same in our body, in which the
lungs, the medullas, or the cerebrum — all of which
possess an undulatory motion — act as the motive force
for all the grosser corporeal parts, preparing the way
for the harmonic flow of the finer tremulations.   This
shows, therefore, that the grosser motions of life consist in grosser tremulations, and that, imitating these,
the finer motions consist in finer tremulations.

CHAPTER III.

§ 1. IT was said that the motion which effects a sensation in our body, consists of a tremulation in the whole system, as well the nervous system as that of the bones, but it cannot be expected that reason will give its consent to this assertion before we have placed before it all the facts from which the connection of all things may be seen. And as Willis, Vieussens, and others, have enriched anatomy with the knowledge of the exact position of every part in the body, we ought now to preface the treatise itself with a short anatomical account, in order to make possible a clearer comprehension of the nature of tremulation.

It should therefore be known that beneath the cerebrum and the cerebellum there is a medullary part called the medulla oblongata, which is provided with many protuberances, glands, and other processes for the distillation of the fluid which is necessary for the membranes, nerves, and bones ; of this fluid more shall be said later on. This medulla is afterwards continued all along the spinal column, down to the legs below ; it runs out of the cranium through an opening called the great foramen of the occiput, entering thence the vertebræ ; from all along this medulla, counting from the cerebrum above, there spring forth, at certain little distances, pairs of nerves which hence proceed to their own functions and uses in the body. Leaving the medulla, these nerves run hither and thither, joining and twisting about one another, again separating in order to join with other nerves, then proceeding further away, splitting up and making greater and greater ramifications, until the ramification itself produces an expanse, which is termed a membrane, or a periosteum, consisting of nothing but the most minute and extreme branches of the nerves.

Beginning from above and counting the nerves in their order as they run out of the two medullas, we find first the *olfactory nerves*, which run directly to the nostrils. Next come the *optic nerves*, which, running out of their thalami, afterwards cross each other, and then again unite, after which each one runs to its own eye. *The third pair* runs to three muscles behind the eyes, giving them power to move and turn. *The fourth pair* proceeds also to the eyes, running obliquely across the other nerves, and making a muscle with a knot or pulley, called *trochlearis*. *The fifth pair* has an extensive ramification ; it runs in four branches to the eyes, and thence to the eyebrows, the forehead, the nose, and the temples ; other branches proceed to the jaws, the lips, the teeth, the palate, and the throat, afterwards sending out a few branches to join with *the sixth pair*, which also runs to certain motor muscles of the eye. *The seventh pair* is called the auditory nerves, which are double, the one soft, the other hard : the former runs to the tympanum, the cochlea, the muscle of the malleus, etc., the latter inclining towards the eighth pair, to the tongue, the pharynx, to certain muscles above the mouth, and around the eyebrows and the forehead, etc. *The eighth pair*, which is called *par vagum*, describes an extensive circuit, running to the muscles of the neck, to the hard branch of the audi-

tory nerve, and to the intercostal nerve ; it then runs
down the body in a straight line, on the way sending
out branches first to the cardiac plexus, whence the
heart receives its nerves and ramifications ; then to
the heart itself and to a number of arteries and veins
around which it weaves itself ; also to the œsophagus,
to the lungs, to the stomach, as well beneath as above,
terminating finally with minute branches in the liver.
The ninth pair runs mostly to the tongue and the palate, and unites afterwards with the next pair. *The
tenth pair* runs simply to the muscles of the neck, but
unites on the way with various other nerves, as with
the great intercostal nerve, and with the spinal nerve.
The eleventh pair goes mostly to the muscles in the
neck and to the diaphragm, uniting finally with the following pair. *The twelfth pair* proceeds to the arms
and to the diaphragm. *The thirteenth and fourteenth*
also to the arms ; *the fifteenth to the twentieth,* etc., run
to the muscles in the arms, the sides, and other parts
of the body, uniting mostly with the intercostal nerve ;
the thirty-first and thirty-second to the testicles ; *the thir-
ty-third* sends a little branch to the penis. *The thirty-
fourth to the thirty-seventh* run only to the muscles in
the legs and the loins ; *the thirty-fourth* to the inguinal
glands ; *the thirty-eighth* unites with a nerve which runs
to the thighs, but runs otherwise to the sphincter muscle of the anus, to the prostate glands, the bladder, the
uterus, and the rectum ; *the thirty-ninth,* like the others,
runs into the intercostal nerve, as also to the crural
nerves, to the muscles of the penis and the anus,
uniting thus with the former ones, and these again
with those that run out of the intestines and the mes-

entery, and these finally with all that system which is produced by the intercostal nerve. Most of the above nerves unite to produce a great nerve, called the *intercostal nerve*, which runs to and from nearly all the parts of the body and effects the principal connection of all the other nerves. The intercostal nerve may therefore be called a connecting sinew and a basis and highway for all the nerves in the body, as may be seen more fully from the whole nervous system, as described and illustrated by the anatomists.

For the further elucidation of tremulation let us consider the most minute parts of the body, and we will see that there is a connection and harmony between all these parts, no matter where they be situated. Take, for instance, any little nerve or point of membrane in the stomach : it will be noticed, first, that both the upper and the lower parts of such a nerve or point are joined to transverse sinews and afterwards are incorporated in the par vagum nerve, which runs to the œsophagus, the lungs, the heart, the pericardium, the throat,— on the way weaving itself about arteries and veins ; it finally flows into the intercostal nerve and now describes another great arch in the body, running to the kidneys, the spleen, the liver, the gall-bladder, and round about the intestines and the mesentaries ; it is thus connected with the branches of all the other nerves which join the intercostal, and is therefore connected also with the legs and the loins, the head, the face, and all the organs of sense, etc. ; finally, by a circuit, it becomes again connected with the par vagum, which terminates in the stomach. Or take any other part of the body, such as the least cuticle on a finger, or on the sole of the foot, and it will be observed similarly that this cuticle, which is nothing but a ramification of nerves, is collected into a greater nerve, and afterwards, by means of other nerves, becomes connected with all other cuticles and membranes in the whole body.

This is, therefore, another clear proof that every sensation is a tremulation in the whole nervous system, and that a sensation is not confined to any particular place beneath the cerebellum, or within any certain protuberance or ventricle here or there, but that it exists, as in one, so in all places simultaneously ; before it can properly be called a sensation, it must have been felt in every part of the body which is reached by a nerve. As soon as the least impression is made on the ramification of any nerve, it is at once communicated to all the nerves and membranes of the whole body, so that one and all must necessarily be touched and impressed. The most direct highway of sensation is the medulla spinalis, which, as to its tunics, is contiguous with the brain ; the sensation or tremulation is therefore at once communicated to the medullary tunics and to the cerebellum. As now all the sinews, threads, or nerves, which are found in any part of the body, originally come from this medulla, it may be understood how the tremulation — as soon as it reaches the fountain or origin of all the nerves — must run like a lightning over them all, in whatever corner of the body they may be. That is, the tremulatory motion is received especially by the medulla spinalis, whence it is communicated to everything contiguous with this medulla, namely, the brain, and all the nerves, mem-

branes, and cuticles in the whole body, all this showing
that a living sensation is not confined to any particular
body of matter or little protuberance in the brain, but
free from any bridle or constraint, it runs like a lightning over the whole body, and then first presents the
sensation.

§ 2. Having now exhibited the connection of all the
nerves, and the harmony of our whole body in respect
to its sinews, it will next be necessary to deal with the
system of the membranes, inasmuch as the tremulation
flows over these up to the cerebrum, using the membranes as a bridge over which it is carried to its
termini above and below.

In general it is to be known that a membrane is
nothing but a contexture of nerves, which by countless
ramifications have been woven into a tissue or coat, to
clothe and distinguish the various parts of the body.
By the microscope one may distinguish innumerable filaments of nerves, entangled and crossing one another;
the lymph flows in some of these filaments, the arterial
or venous blood flows in the others. According to the
latest discoveries, the membranes consist mostly of
nervous lymphatic ducts, the blood vessels constituting
the lesser part: all these, infinitely ramified, produce
an expanse like a retina. Every least part of the body
is thus clothed with a membrane, thick or thin according to the use and nature of the part, and often with
two or three membranes in which the nerves terminate
and thereby compose the chief structure of these membranes. Such is the composition of the stomach, the
mesentery, and everything else, as is held for certain
by some perspicacious scientists.

Chief among all the membranes are those which are called matres or meninges : the upper one, which differs from the other as to thickness, is called *dura* or *crassa mater*; the lower one which is thin and fine, is called *tenuis* or *pia mater*. It is in these that the principal motions of tremulation take place ; in these reside most sensibly the most subtle sensations, and in these, as in little mirrors, may be seen the real nature of tremulation.

The dura mater may be said to produce an expanse over the whole body, for it communicates membranes and tunics to all parts, making a continuous system just as the nerves do. It is expanded over the whole brain, extends itself into all fissures, as between the cerebrum and the cerebellum, and surrounds the medulla spinalis throughout ; all the nerves which flow forth from the medulla similarly cover themselves with a tunic from the dura mater and carry it with themselves to all parts of the body. As now the nerves send out ramifications to all the periostea, to all the muscles, and to all the blood-vessels, it follows that the dura mater provides clothing for all these parts, forming all cuticles, periostea, and integuments in the body, so that the whole contiguous system of membranes is nothing but a continuous extension of the dura mater.

Beneath this latter lies the pia mater which still more encloses all parts which are distinct from one another. It surrounds especially the cerebellum and enters into every sinus there ; it covers all the protuberances and glands of the brain with little membranes, gives a tunic to the ventricles, runs along all the nerves which flow forth from their roots to their foramina in the cranium, accompanies the medulla along the spine as an inner coat, and divides it into two parts. In a word, it encloses inwardly everything that is of a lymphatic or soft nature.

Between the dura and the pia mater the anatomists have also found a thin membrane, called the *arachnoid*. This one encloses the whole medulla spinalis, and, I believe, also the cerebellum ; it lies so close to the pia mater that many may suppose it to be the uppermost tunic of this mater, but it may be seen most distinctly at the punctures whence the nerves come forth from the medulla spinalis. This arachnoid membrane also accompanies the nerves as a second coating, continuing, perhaps, to the smaller ramifications as covering lamina and producing a great number of tunics and periostea. The meninges of the brain, therefore, produce a continuous system of membranes over the whole body, and as this whole system covers a structure of continuous vessels — lymphatic as well as sanguineous — and all filled with their own fluids, it follows that these membranes are hereby kept expanded and distended for the requirements of the tremulation, being more or less attuned, as it were, according to the influx of these fluids. But of this more will be said in another chapter, in which we shall demonstrate the theory of the circulation of the lymph and the blood.

§ 3. It must also be shown how the bones are connected and articulated one with the other, inasmuch as on this connection depends to a very great extent the instantaneous communication of the tremulatory motion. A skeleton clearly demonstrates this connection. It should also be remembered that every bone as to its least part is enveloped by a periosteum, that is, by a membrane extended from the dura mater (although the temporal bone has been excepted by some, yet not by all the anatomists). The dura mater, therefore, applies itself closely to the bones, seeking as it were to incorporate itself with them in order to give them nutriment by its fluid. It also extends little tendons or threads far into the substance of the bones, thus joining the membrane so closely that it can hardly be separated from the bone. In the cranium itself the dura mater applies itself so firmly to the rough surface of the bones, that it is difficult to separate it without causing some injury. By little tubules and points it is also joined to the pericranium so firmly, that nothing is felt in the one which is not also felt in the other membrane. Moreover, in the great foramen, through which the medulla spinalis enters from the head into the vertebræ, the dura mater is so swollen that it seems fibrous and fleshy ; the reason for this is that all the tremulations must flow through this opening, as through a bridge on a violin, before entering the concave part of the cerebrum.

§ 4. From the above we may now have a better comprehension of the communication of tremulation. For it is known that the tremulation flows with the rapidity of lightning over membranes and nerves, from one end to the other, in an instant making the most subtle waves over the whole expanse, like the oscillation in water or in the atmospheres. As now all the membranes are expanses, and are continuous with the dura mater, and as the meninges surround the medulla spinalis, divide it into two, and enclose the whole cerebrum

and cerebellum, it may be seen, in regard to the nature
of tremulation, how quickly this will vibrate from one
terminus to the other. It is so swift, in fact, that we
cannot form a conception of its celerity by any comparison with our divisions of time. It has been shown
further, that all membranes are joined with what is
hard, that is, with bones, from which it follows that
every tremulation in the membranes is at once communicated to the bones; the same motion that begins in
the nervous system is instantly communicated to the
osseous system, thereby assisting the tremulation as
well in respect to the swiftness of its distribution as in
respect to its continuation in the same degree throughout. For if there were no contact with something
hard, the expansion of the membranes would not be
sufficient to effect the communication of the tremulation so swiftly, inasmuch as this motion always loses
somewhat of its force when it meets any thing that is
soft. When we examine the periostea, which surround
the vertebral foramina, it will be seen that these are,
indeed, separate tunics, yet derived finally from the
dura mater, so that there can be no tremulation in the
one which is not ultimated in the other. The tremulation is therefore at once communicated from the spine,
with all its membranes and vertebræ, to the cranium;
the latter, which is the exit and entrance or the very
bridge of all tremulations, is so completely joined and
fixed to the matres, that any motion in one part must
necessarily be felt at once in all the parts of the head.
Moreover, there are concave cartilages on all sides, and
the cranium itself is so porous, that it is especially fitted
for the reception of what is tremulatory. For porosity

contributes more than anything else to the communication of the finer contremiscences; the more porous a thing is, the better does it play with the tremulation of a chord; porous wood, such as cedar or spruce, is far more suitable to conduct the sound of a chord than any harder kind of wood. The all-wise God of nature has therefore created in man a great cavity, surrounded by a very porous cranium, so that it is like the body of an instrument, from which the membranes receive a higher tone or pitch, for the effecting of sensation in the body. And as each one must form his own opinion according to his own brain, we may be allowed to express our belief that the tremulation first runs as far up as to the great foramen of the occiput or the interstice of the medullas, and thence into the cranium to the coronal suture. For the matres are here joined to the cranium, and in the foramen they seek as it were to attune the sound, like a bridge on a musical instrument, whence it flows up and down, and effects in us the quality of the tremulation. *Quod erat demonstrandum.*

CHAPTER IV.

§ 1. IT was said that the membranes, over which
tremulation flows and which carry the motion into the
cranium and over the whole osseous system, are of the
same quality as a musical chord. To originate a sound
there must be a tightly-stretched string no less than a
well-built body of porous wood ; the finest sounding-
boards can effect no sound if the strings are slack.
Thus also with the membranes in our body : such as is
their tension or expansion, such will be the communication of the tremulation to the osseous system, and
such will be the increase of the sensation ; but as soon
as they become slack, they can communicate no tremulation for the production of sensation, even though
the bony system is perfect and in all the strength of
maturity.

It is now our intention to show how the membranes
become expanded and attuned to receive the tremulatory motion by means of the influx of blood and lymph
into their respective vessels. When these fluids press
in with force and swiftness, they cause all the ducts
and minutest vessels to swell up, and as the membranes
consist of nothing but such vessels, it follows that the
whole membranous system then becomes expanded and
distended. On the other hand, when there is present
any force which expels the blood or the lymph from
their vessels, it follows that the membranes are let
loose and become like a slack string which only can
vibrate very slowly and scarcely is able to conduct the

tremulation to the bridge of the instrument. The degree of fulness of life is, therefore, according to the degree of the tension of the meninges : the more these are expanded, the swifter is the course of tremulation, and the greater is the degree of what is called *esprit* and presence of mind; but in the degree that the membranes as it were collapse and perhaps conceal some serum or lymph beneath their folds, in the same degree is the tremulatory motion prevented, and in the same degree do we suffer from absence of mind and of understanding, the body no longer responding to what is quick and prompt.

As now the tension of the meninges is the most necessary condition for any proper kind of tremulation, let us next consider the circulation of the two principal fluids in our bodily system. It is hardly necessary to say anything here as to the circulation of the blood, inasmuch as Anatomy is very rich in knowledge on this subject. The general opinion is, however, that the blood flows from its center, or the heart, through the arteries, which gradually ramify into finer and finer arteries, until they finally terminate in little branches and delicate vessels, to the number of many thousands; all these are distributed not only over all the cuticles, but over all the membranes which are visible in the meninges and in the choroid plexus, until they present a reticulated expanse, figuring in the membranes like the finest ramifications and leaves of a tree. After the blood then has followed the arteries into these least ramifications, it flows into the venules, and thence, by the same pressure, into the larger branches, and so, finally, through the greater veins back again to the heart to repeat the same circulation as before.

While such is the circulation of the blood through the arteries and veins, there are other little ducts, called the lymphatic vessels, which run as it were out of the wall of an artery and across to a vein, like little aqueducts pouring their fluid into the blood which is to return to the heart : these aqueducts are lymphatico-nervous vessels, each surrounded by most minute nerves and membranes. Now, though the microscope has discovered something of this order, still we do not suppose that any one is as yet able to claim a knowledge of the fountains of these rivers. But if we are willing to follow the guidance of sound reason, acting on the suggestion of what has been observed thus far, we cannot but conclude that just as the heart is the propelling organ of the blood, so are the cerebrum and the medulla the fountain of the circulation of other fluids in the body. It is known that the cerebrum and the medulla have a reciprocal and undulatory motion, just as the heart ; this motion must necessarily cause a fluid to be pressed in and out continually, propelling it to the extremities, and thence back to its original fountain. Now, out of the two medullas there flows in the nerves a fluid called the nervous serum, — as will not be denied by any person of common sense, for it is known that the nerves are humid ; the same may be observed in the wood of the hardest tree, in which the sap flows up and down through the pores to all the parts of the tree. These nerves are again ramified into finer and finer branches, until they are finally expanded into membranes which clothe all parts of the body ; these membranes, in their turn, provide a new sheathing for the nerves, and thus accompany them

back to the medullas. The fluid, which has poured itself through the fibrils of the nerves to the extreme
tunics, has thus an area or space in which it may either
unburden itself or else flow back again to the matres
and meninges as to its own source, just as the blood
returns to the heart.

It has been shown by Vieussens that these membranes
extend thousands of threads or lymphatico-nervous tubules on one hand into the bones and on the other
hand into the arachnoid and the pia mater. Through
these the lymphatico-nervous vessels cause to be distilled into the medullas a fluid which that author would
almost acknowledge as an animal spirit; this fluid has
thus flowed out of the medulla through the nerves into
the membranes, and then back again to the medulla,
making a circulation similar to that of the blood. This,
at least, seems credible to us, and it is also possible that
the lymphatic vessels run only from the walls of the
arteries into the walls of the veins, that is, from nervous tunics into other nervous tunics.

As now the blood has its own sources of distillation,
that is, the lungs, the glands, and the chyles, which
form the blood before it comes into circulation in the
arteries and veins, so also does the nervous fluid possess its own glandules, ventricles, and vessels whence it
is distilled in the brain. But let us consider these in
their order. It is known that the dura and pia mater
are covered on the surface by little sanguineous and
lymphatico-nervous vessels; these, again, send little
tubes into the cerebrum, pouring into it the fluid which
the cerebrum afterwards is to distill and pour into the
medulla. The pia mater is therefore, at first, almost

united with the cerebrum ; the upper part of the latter,·
which is ash-colored, is called the cortical or cineritious
part, and consists of little glandules which distill the
fluid ; below this is the medullary part which is whitish
and consists of little *striæ* and tubules, through which
a fluid may be seen flowing out of the cortical part.
The cerebrum is, moreover, divided into two hemis-
pheres, between which lies a hard medulla, called the *cor-
pus callosum* ; all along the latter, towards the ventricles,
there is a soft substance which has been called the *septum
lucidum*.   In each hemisphere there is a ventricle or
cavity surrounded within by the pia mater ; between
the two ventricles there is another called the *fornix*,
and still another, smaller one, has been found between
the cerebrum and the medulla oblongata.   This medulla
arises from four roots and possesses two cornua from the
cortical part of the cerebrum, as also from the cerebellum ; these cornua are called *pedunculi* or processes.   On
the top of the medulla oblongata there is the annular
protuberance, surrounded with its *pia tunica*, and in
connection with it there are two other globules, one on
each side, from which arise the optic nerves, whence
they are called the *thalami* of the optic nerves.   The
cortical part is inmostly in these thalami, but the medullary part is on the surface.   Just below lies the *in-
fundibulum*, also surrounded with the pia mater, and
ending in the *pituitary gland*.   Next follows a number
of other little glands and protuberances, such as the
corpora striata, the *pineal gland*, the *anus*, the *nates*,
testes, corpora pyramidalia and *olivaria*, the *pons Varolii*,
etc., concerning the nature and position of which one
may consult the anatomists.   Below all these the me-

dulla runs through the great foramen of the occiput, and enters the vertebræ, where it becomes known as the medulla spinalis.

Returning now to the nervous fluid, we will understand that it is ever flowing anew from the meninges through the little lymphatico-nervous vessels, which by means of tubules carry their burden into the most minute glands of the cerebrum; from here it flows into the medulla and through the nerves into all the membranes and all the finest and most remote expanses of the body. It thus also flows into the muscles, where the nerves at each point send forth most minute filaments like a fine net or web; these are supposed to close in the veins and to reciprocate themselves through the heart to the finest ramifications of the arteries, and so, in the same manner, to the brain. It is also to be observed that the inmost kernel of the medulla spinalis is a cortical or glandular substance, the other or external part being medullary, consisting of continual striæ which receive the fluid from the membranes and carry it back to the brain, to be distilled over again for new use in the nerves.

It would seem that there is a difference between the lymph which drips forth from the dura mater and that which comes from the pia mater, and that both of these kinds are different from the nervous serum itself. This difference, which is not yet clearly understood, may perhaps be illustrated by the difference in the humors of the eye: the humor which flows between the lamellæ of the tunic of the eye is supposed to be the same with the fluid which is distilled by the dura mater; this humor is aqueous and not especially sensitive, but mixed with urinous matter or with a subtle salt which as yet is but little known to chemistry ; the crystalline humor, which also is the hardest, and which has its own globe or adytum in the eye, seems to be distilled by the pia mater, for the tunics on both sides of this globe are an extension from the pia mater ; nevertheless, as numerous little arteries run into it, and as it has a still finer tunic on the outside, we are unable to make any certain conclusion on this point ; the third humor, which is called the vitreous, and which lies almost on the retina, seems to flow immediately from the optic nerve and its thalamus, and it appears that the nervous serum has considerable affinity with this humor, not only because it possesses an even viscosity, being neither too fluid nor too tenacious or viscid, but because in the most minute networks it makes an expanse by its mucus which flows in through the optic nerve. All this, however, has only been said by way of suggestion.

However this may be, it shows at least that there must be a tension in the membranes if any proper kind of tremulation is to be communicated over the whole of their expanse, so as to effect a sensation in one thing or another. This tension can arise only from the infilling of the vessels, whether it be the blood-vessels or those through which any lymph or juice is circulating. In the blood-vessels there is a certain degree of heat, while the other vessels possess a certain degree of coolness, and both regulate and moderate the natural heat of the living body and produce the proper expansion of all things. Any obstruction in either of the fluids causes an obstruction either in the nerves or in the other vessels in the membranes, and prevents that tension which alone enables the tremulation to present a living sensation.

§ 2. As now the finer degrees of tremulation require an expansion or tension in the membranes, and as the swiftness of the motions and the consequent intensity of the sensation are according to the degree of this tension, it follows that slack or flaccid membranes cannot possibly serve for any subtle activity. In a new-born infant, for instance, everything is still soft and unripe, and there is, therefore, little or no activity; with an adult man, on the other hand, everything has reached its proper expansion, and all tremulations consequently flow promptly and forcibly to their effects, as well in respect to comprehension as in respect to expression; with the very aged, finally, all things must move slowly, and approach more and more to a state of dullness, because with them all the membranes have become slack and wrinkled and receptive only of the coarser undulations, so that they can have hardly any contremiscences in their whole being.

If now, from any accidental cause, the tension has been removed from the membranes, — either by the blood being expelled from the finest arteries or by the nervous fluid being stopped up in the little foramina, — the effect will at once be externally observable: the senses can no longer perform their functions, and the thought and the memory no longer remain distinct, but the man becomes like a mere form, almost void of life, the vital fire being gradually quenched and approaching a state of quiescence or death. This may be illustrated by the conditions of the body during various states of passion and affection.

A sudden fear causes the blood to rush back to the heart in an instant: it fills up the greater veins, withdraws from the finer arteries, and completely exhausts the most minute vessels; the muscles are deprived of blood; pallor covers all the extremities; the membranes also become exsanguious, they lose their tension, lie down slack, and become altogether unfit for the reception of a tremulation. Hence each one of the senses is deprived of the greater part of its sensitive power; the eye loses its acumen, and the same happens to the ear and the other organs; the thought and the imagination become indistinct, and the life is in danger, nay, is sometimes extinguished before the blood has been able to force its way back to lift up the collapsed vessels. Sometimes there follows a tremor, a quivering, a stroke or convulsion throughout the body, for the greater part of the life is lost as soon as the tremulation no longer can flow over a stiff expanse.

Amazement produces similar effects, in so far that everything in the body then displays a tendency to come to a standstill; even the involuntary motion seems to have stopped; the blood has hardly any impulse toward a new circulation; a general state of forgetfulness and stupidity results, with a limpness and placidity in all the membranes, until finally a more full tremulation is able to pass over them.

Swooning is also of a similar nature : through a sudden alteration the nervous fluid or the blood rushes out of the membranes or else becomes obstructed in one place or another so as to be prevented from flowing forth to its expanses, networks, and membranous plexuses; the latter, therefore, collapse at once and become unable to receive any further tremulations, and the subject remains as it were half dead, until the blood can again flow into the membranes and expand them as before, when the life finally regains its tremulation and the tremulation its life.

In *paralysis* or other *convulsions* it is known that something has closed the road to the membranes or nerves through which the fluid must pass, and consequently has obstructed the motion which ought to glide over these as the proper bridges for communication in the body. Such strokes and obstructions show conclusively that there is a real circulation of the lymph, for if the fluid is obstructed in the nerves, then their membranes no longer receive any of that lymph which must flow out to the extremities of the nerves, but they lose their tension, and the tremulation comes to a stop at the very beginning of its course without producing any sensation in the entire half of the body.

In the case of those who have died of *apoplexy*, or have lost the real acumen of their senses through a wild or extraordinary tremulation in the bones and the cranium, it has been observed, when the skulls of such persons have been opened, that the dura mater has been bloodless and slack, sometimes twisted into folds and wrinkles, and sometimes with the lymph exhausted between the dura and the pia mater ; the cortical and medullary parts have then been found soft and watery, the glands and the pituitary body distended, the ventricles filled with viscosity, the medulla spinalis quite rheumy and as it were inundated by stagnant water ; many such cases will be found described by those who have investigated and made notes of such things. All this goes to show that tremulation is a *fac-totum* as to everything living in our body, for as soon as the little membranes are no longer in a state of tension, but folded and slack, we lose all that is initiative, that is, we lose our senses, our thought, and everything else that makes us truly living.   On the other hand, it is wonderful to observe how everything gains a new life, each sense a new presence, and each different sensation a new alertness, as soon as a new circulation takes place, by which the fluids may again flow forth to their extremities to expand each minute vessel and hence the whole expanse or membrane which is woven together of such vessels.

In a condition of *courage*, for instance, the blood flows freely forth into all its arterial vessels until it reaches the veins; it is then impelled into the very cuticles, and spreads a blood-red or crimson color over all the tunics; the two matres, also, have now become full of fluid and thus expanded, so that the tremulation is able freely to play over them in swift motions; all the senses possess their proper life, and everything has its proper termination, comprehension, and presence.

If by any other passion, such as that of *Amor and Venus*, the blood has been driven into the extremes of the membranes, it will again be observed that all the little membranes have become expanded and prepared for tremulation to put vigorous life into all the sensations.   A condition of *anger* causes too hasty a circulation of the blood, fills up the little vessels too abundantly, and distends the canaliculi too greatly.   The same is the case in a state of *heat*, as in a sickness or fever: the smallest vessels swell up into bladders, as it were, driving the internal fire into the very cuticles.

Thus also in *drunkenness*, when a person has too freely imbibed such fluids as make the blood volatile : the blood then presses forward so as to expand and swell the sanguineous vessels too strongly ; each sense then reaches the highest degree of its life ; the tremulation becomes uncontrollable, making wild and disorderly movements in place of the ordinary and even ones. If the quality of the membranes be then examined, it will be noticed that they are gradually losing their evenness or smoothness, the sanguineous ducts make tumors here and there, and cover the membranes with waves, as it were, whence the tremulation receives a different quality, and the arteries undulate from the excited blood within. A tremulation may then, indeed, pass over the cuticle and its nervous system, but, since the arteries are thus distended, a softness will be in the road here and there, preventing the proper play of the tremulation, and obstructing the lymphatic ducts through which the tremulation should be communicated in the first instance ; the motion is thus turned into a different kind of tremor, which does not correspond with the usual one ; it communicates a dull and stupid tremulation in place of the proper one, and causes what is known as madness.

§ 3. What great changes do not instantly result from any accidental injury to the dura mater ! Convulsions, swooning, and general collapse follow quickly, and apoplexy often leads to death ; in a word, it is the shortest road to the stoppage of our whole moving life. If the medulla spinalis or the cerebellum is pierced by the smallest point of a needle, we know that death follows immediately, sometimes by a lymph being emptied out

of its vessels between the membranes or the skin,
whence the matres at once become wrinkled and laid
into softer folds, to the hindrance of the tremulation
which should pass over them.   All this shows that our
proper life resides especially in the membranes, and
accommodates itself to their state of tension or slack-
ness ; the tremulation, therefore, accommodates itself
similarly, carrying the life into whatever degree that is
permitted by the tension of the membranes.

While studying the nature of the membranes and
their expansion, it seems difficult to form any other
theory than the one here indicated.   Now if our life
and nature were to consist of something else but those
subtle motions which are called tremulations, what then
is to be thought of those cases in which men have lost
a part of the cortical substance of the brain, and yet
have remained in possession of their senses?   Cases
have been known in which parts of the cranium have
been removed by trepanning, when it has been found
that the cortical substance has been evacuated by hand-
fuls ; other cases are recorded, in which both the medullary and the cortical parts have turned into a purely
watery liquid, in which all the formerly distinct parts
have been swallowed up, so that no proper glands, protuberances, or ventricles remain, but the whole has become like a slough of water ; and yet, notwithstand-
ing all this, the senses have remained in their natural
condition.   We are therefore forced to conclude that
life resides in the ultimates and thus in the meninges,
and that it will remain there, let the internal parts be
as they may, provided they are still able to furnish some
kind of an even and proper fluid for the nerves, whereby

the membranes may be kept in some state of tension.

Our theory may be seen illustrated still further in those cases where the whole brain has been ossified or petrified. There are on record at least two such cases, one of which may be seen in Professor Dr. Bromell's * beautiful collection. In such cases, it would seem, the fluid must be distilled immediately from the matres, dripping by little pores or canaliculi through the inner, petrified part, as in the case of other bones. It cannot, therefore, be denied but that the surrounding membranes really effect that which is usually ascribed to the inner part, that is, a motion, which is communicated to the whole system of nerves and bones, and consequently to the whole body, for the effectuation of all our sensations.

If a subject of the vegetable kingdom may be compared with an animated thing, it will be seen that nature has a similar character in both kingdoms, that is, that the living and growing force depends upon the coats and the bark through which it is flowing. Take, for instance, a tree, the trunk of which lies broken on the ground : as long as the bark still joins the two parts of the trunk, the leaves and the fruit will still continue to live on the tree; but if merely a strip of the bark be peeled off round about the trunk, all the verdure and growth will vanish and the whole tree will die. From this comparison, therefore, it is evident that it is the pleasure of nature to place what is chief and most noble in what is most ultimate, and that all life must depend on this ultimate.

* Dr. Magnus Bromell, professor of medicine at Upsala, afterwards President of the College of Medicine in Stockholm, and Physician to the Royal family. † 1731. Tr.

Now it may, indeed, be objected, that the dura mater is often injured in cases of trepanning or through other causes, when yet the senses are improved and regain their proper order; still, when this happens, it will be found that the injury has been received in places which are not especially connected with the general tremulation, for if the membrane over some particular nerve be injured, the collapse does not extend itself further than to the extremity of that nerve. Similarly, if an incision be made in the meninx opposite a suture, the proper tremulation is not thereby altogether lost, although it is made somewhat slower; but if injury be done to such a part of the dura mater as may be supposed to be the very bridge or focus of the tremulations, either on the cerebellum or on the medulla spinalis, then there will at once result a shivering or twitching over the whole expanse, or the vessels will be emptied of the lymph, so that the tremulation can no longer pass over its usual bridge except as over a slack string.

§ 4. As has been said, the tremulation requires a tension for its swift and proper communication to the cranium and the other bones, in order to be felt in the organs of sense. This may be illustrated by the quality of a musical chord, which, if slack, will merely undulate slowly without producing any sound, but, if drawn tight, will gradually give a sound which becomes clearer and sharper in the degree that the chord is drawn toward the bridge. It is the same with a drum: if the membranes are slack, the tremulations hardly reach the hundredth part of their proper distance, but if drawn tight, the swiftness of the tremulation and consequently the strength of the sound are increased so as to reach great distances.

Looking further into the cause of this, we will find it illustrated by the pendulum : the longer the pendulum, the longer time does it require for each vibration, but the shorter it is, the swifter is the vibration within the same measure of time. Instead of a pendulum take a rope, or line, which naturally hangs in a curve ; if now

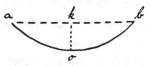

the line a o b is swung to and fro, the times required by the vibrations of the point o will be equal to the times required by a pendulum of the length of o k. If the line is drawn tighter, the cross-line k o will at once become shorter, like a shorter pendulum, resulting in a swifter vibration at o, and the tighter the line is drawn, or the more nearly into a straight line, the swifter are the vibrations. The idea of the geometricians is, therefore, that the times are as the radii and the length as the squares, which is called a duplicate ratio.* If, then, a musical chord or string always hangs more or less in a curve, it is but natural that the more it is drawn tight, the more does it approach a straight line, that is, the shorter the pendulum k o, the swifter the vibration.

It is the same with a membrane : the more it is expanded or stretched, the more are all parts of it expanded or drawn tight into straight lines or a straight surface, and the more swiftly can the tremulation pass over this surface ; thus also with a string, to which a ball is suspended and swinging, if we take hold further

---

* We understand this to mean that the length of the pendulum is as the square of the time of vibration. Tr.

down the string, the vibration will at once hurry on faster; a musical chord, when gradually drawn tight, will at first give a coarse sound, then a finer and finer as it is drawn tighter and tighter. A tremulation in the body possesses exactly the same property, requiring more or less of tension if it is to express properly any sound or sensation.

§ 5. There exist with men many different kinds of genius or temperament, which arise simply from a difference in the nature of the expansion of their membranes. Those are said to be of a *sanguine* temperament, with whom the blood is thin and volatile, so that it flows like an ether into the least vessels and inflates them in an instant; these persons above all others possess membranes that are filled with blood and are stiff for the reception of tremulatory motions; all things with them are movable in the first instant and ready for effectuation in the same moment; they are more inventive, more communicative than any other persons, and more inclined to anything that requires blood in the membranes.

Those are said to be of a *melancholy* temperament, with whom there is a thicker kind of blood which flows with greater difficulty into the most minute of the vessels and hence does not effect its circulation in as volatile a manner; their membranes are expanded only by the greater vessels, and the matres are somewhat slack and less even or smooth than with the former, so that everything with them is effected more slowly. *Phlegmatic* persons, also, have a slow life, because a lymph or serum seems to predominate in the membranes, the blood having less space or lodging, whence there is less heat in the body. [Certain words in the sentence next following are obliterated in the manuscript, making four lines unintelligible.]

. . . Now, as a mass of ramifications of the nerves weave themselves about the arteries and the veins,— not only in the body itself, but also in the brain,— composing what is called the nervous tunic, it follows that as soon as any passion has originated, the blood is more or less under the control of the nerves; by the contraction of the nerves the blood is closed off from its finer vessels, while by the expansion of the nerves the blood is permitted to flow freely or is propelled forward with increased pressure, so as to expand the membranes. From this cause comes that immutable law which is exhibited in the membranes. For if the blood is obstructed in the membranes, there results at once a different attuning of the whole nature of man.

CHAPTER V.

THAT life is tremulation, or that whatever is living in us must be expressed by the motions of tremulation, is manifest not only from the connection of every least part, but especially from this fact, that life has a tendency to accommodate itself altogether to the solidity in the systems of the body. For it was shown above that tremulation requires not only a tension, but also what is hard or solid, in order to effect an intensification of the sensation as well as a communication, just as a musical chord requires a hard and concave body, which effects the reverberation of the sound as well as the aptitude of the chord for tremulation; otherwise the vibration of the chord would hardly reach the tympanum of the ear, but would vanish in the folds of the external organ. It is hardness or solidity, therefore, that contributes a higher degree of the perception of tremulation, or which makes us conscious of the sensation, as may be seen from the whole life of man, in all its ages.

Consider the state of a *new-born infant*: there is nothing fixed or stable in his whole body, no firmness in any bone; the cranium bends to the touch of the finger; there is no consistency in shoulder or leg; the dura mater has not yet become fixed to its sutures; that which is to become hard is still a serum or a yielding substance, so that no tremulation is as yet able to pass from what is soft to what is hard, or vice versa, to effect a sensation; the meninges of the head lie on a very soft cerebrum ; the arteries or other vessels have not yet fixed themselves firmly to their canaliculi so as to prepare the membranes for the reception of tremulation ; the cerebrum itself is not able to contribute any expansion or tension to its tunics, but must, like a mere fluid, leave with the matres whatever impression is made upon them. Here we may see, as in a picture, how life, and the use of the senses, must accommodate itself to what is hard, and how all the tremulations are kept back on account of the absence of this hardness. The external senses are not yet fully alive with the new-born infant ; he can perceive nothing distinctly ; hearing, sight, and all the rest, are to him like a shadow or a cloud, in which nothing distinctive is possible. The cranium being soft, the membranes have not as yet gained any proper expansion, have not been attuned, as it were, but are like loose strings, over which a tremulation may indeed pass, but only with an undulatory and dull motion, without producing distinct sounds, and without power to reach the bridge of the instrument by any other natural motion than the involuntary one.

But the compositions in the body soon begin to gain stability, fixedness, and expansion ; the sutures of the skull are gradually closed and knit together, thus drawing the membranes toward the cranium ; the nerves also grow harder because the medullary or striated part in the greater nerves is drawn into hard filaments on which the membranes of the nerves lie as on hard bottoms. Life then begins to become properly living, the senses receive their alertness and acumen, everything gains more and more knowledge of its own use and quality. In a word, nature then begins to express

itself, the sensories find their termini, and the tremulation gains freedom to pass from ultimate things to in-
mosts.

Passing by the years of growth, when the various
bodily systems are continually adding something to their
size and hardness, we arrive at *Adult Age*, when the
efflux of the nervous fluid becomes proportioned to the
expanse of the body, so as to give it sufficient nutriment, or when the medulla is fully able to support the
full-grown bones and body.   Our whole system of bones
has then become fixed and has gained its proper dimen-
sions and hardness; everything is firm and living; all
the senses are in the fulness of their uses, and the internal senses, such as the memory and the thought,
are at their highest point of development.   The membranes, also, are then most firmly expanded by their
vessels, such as the arteries, veins, and lymphatic ducts ;
the medulla has been shaped into long striæ ; the nerves
have become firm ; the bones and the cranium have
fully developed their porosity ; and the lamina of the
bones have gained their hardness, so that tremulations
may be properly received.   Since now it may be seen
that the tremulatory motions effect a better sound or
sensation by means of tension or hardness than by the
contrary, we have also proved that life, or what is properly living in us — which is a distinct perception or
discrimination of all things — consists in tremulations.
For when the compaginations or systems, over which
the tremulation is to pass, are out of their order and
use through any improper softness, then also is all perception suffering, but as soon as the frame-work is
ready, and the whole key-board furnished with taut

strings, then only is it able to convey the sound or the perception, which propels itself by means of tremulations.

Now when *Old Age* comes on and inclines toward the end of life, the condition of the membranes and of the whole frame is equally notable. The medullary substance, both in the spine and in the cerebrum, is then becoming more and more hard and sinewy, for its whole tendency is to run into sinews. It has also been found that the medulla is more empty of its fluid with the aged than with the young, whence, with the former, the nerves no longer receive their proper nutriment ; the fluid flows more sparingly to the membranes and the extremities ; the serum and the lymph become more and more dried up, making the matres more and more slack, wrinkled, and folded ; the finest bloodvessels can no longer penetrate to the surface, but are closed off by the hardening external forms, are kept away from their most subtle ducts, and thus cause the meninges and all the other cuticles to fall into folds and wrinkles ; everything is becoming more and more empty and light, and man is failing as to every part. It may thus be seen how the tremulation is as it were shut off from effecting the life of the senses ; the membranes, over which the motion is to pass, have become displaced, loosened, and slackened ; the tremulation must needs stop at the first initiative, or it runs forward in a dull manner, almost without any sound ; the acuteness of the sight as well as of the hearing decreases and becomes obscure, the membranes of the eye contracting to the injury of their convexity. Everything is more languid, slow, and tepid than before. The internal senses are

similarly decreasing, accommodating themselves exactly to the condition of the membranes which now
are slack and wrinkled; all this a proof that the veri-
est life of man begins to fade away as soon as the
tremulation is in any manner prevented.

Compare now what has been said above with the life
of the higher animals, and it will be seen still more
plainly that tremulation makes the greater part of our
living force, and, in fact, takes the role of nature itself
in our life.   Such animals as are born with the bony
system perfectly developed, or nearly so, come quickly
into the use of their full nature, as a native or innate
thing.   A chicken, for instance, possesses most of the
consistency of its whole skeleton when it is first hatched
out of the egg; it can walk, it can see, it can hear, and
move its wings and neck; in a word, everything within
it is at once attuned for the reception of tremulation.
There is, therefore, nothing dull with this little crea-
ture, but it runs at once into the full enjoyment of
its very nature, without having to be nursed for any
length of time.   Of the greater animals, some attain
their full growth in a few weeks, some in a few months,
the largest in three or four years, after which their
senses are developed as perfectly as with us when we
have reached adult age.   Everywhere it may be seen
how nature accommodates itself to the firmness or hardness of the nerves and the bones.

Now if the life of the senses were effected by any
other means than tremulation — by some volatile force,
for instance — why is it that this must still be connected with a hardness in the bones and the membranes, as a necessary condition?   Could not this force

flow through a soft substance, or a liquid, just as well, or better, than through what is hard? Why must the hardness open the door and prepare the way? Or why should not the same way be just as open with the aged as with the young? Is it not clear that much traversing must open the channels wider and wider, and that, therefore, according to such an hypothesis, all the senses must be more wide awake with the very aged? But as it has been shown that everything of life is accommodated to the stiffness and tension of the membranes and to the hardness of the bones, we cannot form any other conclusion than that the sensories, which are analogous to these parts, depend upon the tremulations or motions upon these parts, and that the tremulatory motion, by this means, displays its real nature in its greater or lesser degrees.

But lest any one should think that the senses themselves might be fully developed with the new-born infant, and that he is simply unable to give an expression to his sensations, we will point to this wonderful fact, that the whole of that period during which the meninges, with all the other coats and membranes, are soft, is a period of oblivion, from which the infant is utterly unable to retain consciously the least impression for a subsequent period of life. Whatever may have been effected by the senses during infancy, seems to have been erased with the youth ; the most important impressions and experiences, and the most common habits of life, received and attracted during infancy, and afterwards retained in the organs of the nerves as the very nature of the child — all this remains in the memory less consciously than a dream, although, as was said, nature still

retains it in the sensories as the initiament of life.   As
an illustration let us consider only the faculty of speech.
During the period of oblivion the organism of the
mouth has been taught by a variety of things how to
strain forth the sounds by different vibrations, and how
to articulate the words ; everything that the nurse has
done for the infant remains with him, but the first efforts and the original habituations are afterwards forgot-
ten.   That which became a habit has now become
nature itself, and it effects the speech in a moment
without the least trace of the original effort.   We may
therefore conclude that all such things consist of most
subtle tremulations which derive their life from the
hardness and tension of the vessels and organs ; but
as soon as these qualities are lost, the recollection begins to vanish, as with the very aged and decrepit, and
with others in whom the dura mater has become greatly
disordered.

It is, indeed, most wonderful that a man excels all
animals chiefly in this, that he reaches his maturity later
than they, and that the very thing which may be ac-
counted his imperfection is really the chief means for
his perfection, and for his exaltation above all animals.
Besides being gifted with a Soul in a reasonable understanding, we have been so ordered by God that all our
membranes and bones must require a long period in
order to become fixed and hard.   In the meantime all
our organs are disposed for the reception of ever new
impressions, all of which require their time, until we
have reached adult age.   Hence also we may see what an
advantage there is in our approaching slowly to our maturity, namely, that the understanding is able to increase

and be more and more perfected, so as finally to present a man who can exhibit a ripe understanding, built up by a multitude of impressions and experiences. An animal, on the other hand, attains the full nature of its parents after a very short period; its membranes and bones become fixed in that nature which has been produced by the impressions of its entire breed or race, and there is soon an end to all further increase.

I know, however, that many may make this objection, that even those membranes which cover the soft parts, and which are not expanded like drums, are still able to cause sensations and to transmit tremulations; and that a tremulation in the bones can effect the same, when yet these possess nothing sensitive. But to this objection I would answer, that I do not at all mean to say that tremulation is distributed over a membrane merely as a membrane, but only in the degree that the membrane is expanded by a fluid; and, further, that a tremulation in the bones does not effect a sensation merely as a movement in the bones, but simply that the latter communicate a tension and a tightening to the periosteum and the membranes, and these, consequently, to the lymphatic canaliculi. For the first tremulation strikes the fluid part, the contiguity of which causes the motion to run over all that is fluid within all the membranes and meninges; and as there are continuous valvuli, it follows that the tremulation, while on the way, strikes against all of these, whence it is carried into the membranes, and so from the latter into the bones, in which a stiffness and hardness are necessarily required. For if there is nothing hard to contribute a tension to the membranes, it follows that the latter must remain slack and unstrung, whence the tremulation cannot run through the lymphatic canaliculi in any proper manner, but must pull and twist the membranes, to the hindrance of the communication. It may be seen from this what the hardness in the bones and the tension in the membranes contribute to the liveliness of the motion in the lymphatic system.

CHAPTER VI.

§ 1. IT is in the sense of hearing, above all other senses, that we may most advantageously observe the real nature of tremulation. All the organs and membranes connected with this sense are so prepared for the reception of this motion, and everything is so formed and arranged with cavities, bones, and turbinated, spiral passages, that the tremulation is excellently fitted to produce and communicate a sound or sensation in all that system which is included by the nerves and the bones.

Considering the anatomy of the organ of hearing, it will be observed, first, that nature has here created a channel through which the tremulatory air must flow, that is, in the curvatures of the external ear, which are able to gather in and concentrate a great deal of the tremulatory motion in order to bring about the better cooperation and cotremulation in the internal organ. Further in, this concentrated air is received by a bony canal, across which is placed a stiff membrane, known as the tympanum, or drum of the ear; the concentrated tremulations then strike against this membrane, effecting a vibration of the same kind as that which had taken place in the air outside the ear. Immediately within, or on the other side of this membrane, there is a cavity or *meatus tympani*, in which are placed the little bony bodies called the *malleus*, the *incus*, and the *stapes*, each connected with one another by articulated joints, so that the motion and tremulation of the one must be communicated to the others. It is this *mal-leus*, or little hammer, which especially causes the tympanum to bend inward or outward, making the membrane stiff or slack, or more or less springy, as may be required by the necessary degree of the attention, or by the strength or softness of the sound.

The *stapes*, or stirrup, lies in a special space, by itself, together with two minute membranes expanded like a little drum, so that every vibration in the bones of the stapes is instantly communicated to this drum. For the further promotion of tremulatory sound there is also a labyrinth, or bony and turbinated passage, formed like a sea-shell, according to the nature of sound. The widest part of this passage, next to the base of the stapes, is called the vestibule, which also has a membrane of its own, of an oval figure, as its base. Up to this vestibule there run ten or twelve little foramina, through which run little threads or nervules, a part of which make the periosteum, the other part running on through the cochlea round about its walls ; these nerves appropriate the sound and communicate it to the other periostea and membranes which are of the same root or contiguity. Behind the other part of the labyrinth, and opening into the vestibule, there are certain small, semicircular canals, bent about like horns or trumpets, which still more augment and concentrate the sound toward the interior part of the cochlea, after which the tremulation flows on into other cavities and membranes.

From the whole anatomy of this organ we may thus know what kind of a mechanism the tremulation requires for its proper communication to the membranes, nerves, and bones in the body. The most important thing to observe is that the little membranes keep close to what is hard, and stand expanded over the air in the concavities like little drums, whence the tremulation is communicated to other membranes ; further, that the membranes weave themselves like the finest network about the bones, like musical strings over the bridge, and, like the bridge, is fixed to the solid part of the instrument into which the tremulation must be carried, if there is to be any strength in the sound. The chief membrane in the mechanism of the ear is the one against which the tremulation strikes in the first instance, and injury to this membrane means injury to the whole organ of hearing. It will be observed that this membrane is joined to what is hard, and that it is derived from the periosteum of the skull; some suppose that one of its lamellæ (for the tympanum itself is said to be double) is an extension of the dura mater itself, although, indeed, the periosteum of the skull has so close a communication with the dura mater that it may almost be counted as contiguous with it ; we may thus see how this membrane connects itself with all that is hard round about. Within the cavity of the tympanum we find all the little bony bodies, the hammer, the anvil, etc., all of which appropriate the same tremulation and propel it further inward. The further it penetrates, the more little membranes does it meet ; all of these are joined to their own bones as their periostea, which afterwards spread out into an expanse, and thus each membrane is endowed with its own solidity and hardness, as a bridge to which the tremulation may be carried. Beneath the stapes, and within the labyrinth, nay, everywhere within the petrous bone,

there are wonderfully constructed cavities and cartilages, all surrounded with periostea and meninges, here
and there extended over foramina, all of these being
little harmonic skins or strings by which the sound is
carried into what is solid.   The most subtle mechanism
in the construction of the ear consists, therefore, in
this, that all the membranes are periostea, so that these
may be able to communicate their tremulation to the
bones, and thus multiply and distribute the vibration.

§ 2. The nature of tremulation, as shown in the
above, may be illustrated still further by many of the
experiments which have been made in the art of music.
It does, indeed, seem wonderful that so small a membrane as the one which closes the auditory meatus and
which is expanded over the little pores and entrances
to the bones and cartilages, is able to effect a tremulation throughout the entire solid system of the body,
and that so small a fountain can produce so great a
motion.   But such is the nature which is peculiar to
tremulation.   What a commotion and distribution of
sound is not. caused by the membranes of a battle-drum
or kettle-drum, or by anything in which a membrane
is expanded over a solid body!   What a difference
would there not be if the membranes were expanded
over a soft substance!   They might be expanded to
their utmost capacity, and still give but a dull sound,
unless their edges were fixed to a hard or solid bottom.
It is, therefore, the hardness alone, which, by its corresponding vibration, can contribute strength and distribution to the tremulation.   Other vocal instruments
show this still more plainly.   The most rudimentary
kind of a violin must have its bridge and must have its

strings fixed at both ends to a solid body, in order that the corresponding tremulation in the body of the instrument may make the sound sharper, stronger, and more continuous.

Tremulation has this further peculiarity, that the least of its motions has the least regard for the mass or volume of the body in which it moves; it regards neither weight, nor hardness, nor grossness, but the contremiscences, as soon as they have originated, run like lightnings over the whole of that body which is subjected to them, while local motions, on the contrary, show respect and fear, as it were, for heavy bodies. This is again illustrated in musical instruments. A string in a piano, when touched, at once sounds and plays its vibration over all that solid part with which it has a contiguity, thus permitting the tremulation to carry its reverberation to greater distances.

From this, also, we may see how small a cause is needed to produce too wide a distribution of the sound. Anything touching the body of an instrument, be it only by a single point of contact, is at once subjected to the tremulation in the instrument itself. If the instrument were in contact with a mast or any long pole, and if we were to put our ear to the other end, or touch it with our teeth, we would both feel and hear the tremulation which has been caused by a slight touch on the little string in the instrument. The tremulation is often endowed with a special sound from the nature of the place in which the instrument is; if there is a great rock below, one may at once notice that the sound is affected by the tremulation in the rock, something which deadens the usual sound in the bottom of the instrument ; if there is a cupboard, box, or case in the room above the piano, the resonance must necessarily be increased hereby.   All this shows that everything contiguous is set into a corresponding tremulation by the touch on the little string.

The increase of the sound is also affected by the quality of the material which constitutes the bottom of the instrument ; it is of one kind if the instrument is made of oak, another if made of spruce, cedar, etc. ; the lighter and more porous the material, the heavier is the reverberation and the cotremulation.   The tremulation is also subjected to great variations according to the varying thickness of the bottom of the instrument, or according to the hardness or softness of the wood, or its dryness and brittleness, or if it is close to some metallic object, or if the string is wound with horsehair or any other soft substance ; all this a clear proof that the smallest membrane or string, which is fixed to a solid substance by the two ends, is able to effect a corresponding tremulation in the most massive objects and to communicate the tremulatory motion to everything that is contiguous with it round about.   If I should add Chapter VIII. to the present copy, you would see, from what I say there in respect to the distribution of tremulation, that the most minute vibration is able to permeate the greatest of bodies, even as the least contremiscence of a violin permeates the whole room where the music is performed.   This is proved incontrovertibly by our own sensation, inasmuch as our hand can sensibly receive the tremulation by touching the wall of the room or the body of the instrument.

A few palpable illustrations may show still more plainly the reason why so small a cause can produce so great a flow of tremulations. In the shafts of mines one will see hanging great cables, weighted with hundreds of pounds; any small motion at the upper end, such as pulling it with the hand or striking it with a stick, will cause the whole cable to vibrate from one end to the other, the tremulation flowing or undulating up and down with swift, serpentine motions, such as may also be seen in the air or the water; the weight at the lower end is thereby lifted up and down, often with such a force as to cause danger to any one standing near, and yet the whole motion may have been produced by a rather insignificant cause. It may thus be seen that the tremulation or vibration has no regard for the weight below, and that the force increases mechanically according to the length or distance of the cable. Again, if a long rope is held by two persons, and one of them pulls quickly at one end, even though with only a finger, the other person will experience difficulty in retaining the other end. Or still better, a long rope, held horizontally, naturally hangs in something of a curve, similar to a parabola. Now in order to stretch this rope into a straight line, one needs a strong 'machine, and yet it will be impossible to stretch it so stiff that there will not always remain some invisible curve according to the length of the rope; a tremulation, passing along the rope, makes a continuous series of such curves, and as each one of these possesses its own mechanical force which counteracts the effort to stretch it into a straight line, and as these curves are swiftly passing on to the other end of the rope, we may not wonder that the impulse and the tremulation has increased in force when reaching its extremity. All this is one with the phenomenon which we observe in the little membranes of the body or the strings of the instrument, which when affixed to their solidities, give so great a force to the tremulation, that the hardest substance must partake of the tremulation, no matter what its size may be.

§ 3. We may now be able to understand more clearly wherein the sensory in the ear consists, and how the tremulation is able to distribute its motion over the entire osseous and nervous systems from so small a beginning; all the membranes, which are intended for the distribution of the motion, are attuned and affixed to what is hard; the first, which is the membrane of the tympanum, is contiguous with the pericranium, so that the least tremulation in the tympanum effects a corresponding tremulation in the cranium and in the petrous bone; further in we find certain membranes which extend from the periostea over little cavities such as may be seen in the *stapes*, in the *fenestra ovalis* and *rotunda*, and in the vestibule of the labyrinth, all of these being little tremulatory drums and cotremulating cartilages; still more interiorly we find wonderfully constructed rooms of cartilage and bone, all covered with little membranes which contribute their share to the tremulation of the entire systems; the use of the cavities is to carry the motion over to the other side and to effect a cotremulation in various places, so that the whole motion may spread with force over the membranes of the brain, the lymphatic canaliculi, the teguments of the nerves, and the bones. The sensory itself consists, therefore, probably in this, that the vibration presses in with force upon everything bony and membranous in the body, whence the communication of the tremulation throughout the system effects the sensation, and the sensation effects that which is perceived by us as a sound.

§ 4. The same testimony is given by those sounds which enter the systems by ways different from those of the usual organs of hearing. First of all, it is to be observed that the sound itself does not reside in the membranes of the tympanum, alone, but rather in the interior membranes, and especially in the dura mater and in the tremulation in the solid parts, and this without regard to the means or methods by which the motion has entered into the body. Any tearing of the dura mater, which may result from a blow on the head or from too great an exertion, is followed by a crashing sound or report as loud as if the head were between two cannons when being fired. When, therefore, the sound does not flow the usual way, nor touches the ordinary membranes, but originates further in the interior and thence sends a tremulation into the cartilages, bones, and nerves and membranes in the body — when we then seek to discover where the tearing or breakage has taken place, or where this interior sound has originated, we may find the cause far beyond the whole auditory canal ; this has been established by the science of anatomy, from numerous cases. The sensory itself may therefore reside in whatever part of the cranium it pleases, provided only the tremulation is distributed over the same systems that are affected by the usual sound from without, through the external organ.

There are many other phenomena which show that

the real sensory consists in the tremulation of the cranium. During a dream, for instance, we often carry on long conversations with imaginary persons, or we may hear whole melodies or other sounds which affect us exactly as those sounds which enter by the external way ; when recollecting the dream during the following day, it seems altogether as if we had heard actual sounds. It is well known that the cerebellum (*hjernan*), while we are sleeping, is in a state of perpetual agitation ; everything is in the effort to react and to restore itself to the proper order for the reception of a new motion and activity ; an uninterrupted stream of tremulations is then flowing over all the systems, expanding everything and filling it anew with blood and fluids, mending, correcting, and restoring all that which has fallen into disorder by disharmonic tremulations during the preceding day. This, therefore, is a clear proof that sound is really an internal tremulation in the cranium and its membranes, and that it does not exist only in the little membranes of the cochlea and the labyrinth.

In fantastic imaginations, also, persons are able to hear various sounds and connected conversations, so that they sometimes persuade themselves that a spirit is speaking with them. I have spoken with a woman, who every day continually heard the singing of hymns within her, from the first to the last verses ; these hymns were often such as she herself had never heard or sung ; she diligently sought help and cure from clergymen and others, but in vain, for the melodies and songs continued in the brain as if she were perpetually attending a great concert. Does not this show that the

sensory of hearing consists essentially in internal tremulations in the cranium, which are able to extend themselves to the ordinary organ of hearing if only there
be a similar motion in the matres. A singing or ringing sound is also noticeable in the ear, when the matres
or membranes are diffused and greatly distended with
blood by the arteries, whence the tremulation is unduly
hastened over everything contiguous.

The contrary takes place when the internal membranes become more thick or slack, as when the arteries
are dry or when the blood has escaped from the most
minute vessels, thereby causing a pallor on the surface,
as takes place in a state of fear and by various accidents
and diseases, when neither the external nor the internal
membranes are able to receive the tremulation and still
less to carry it round about. For a slackening of the
membranes must necessarily cause a letting back or
regurgitation in all the lymphatic vessels, so that when
a tremulation then enters the most minute of the canals,
it finds the membranes slack and wrinkled and thus obstructing the communication.

Other phenomena show still more plainly that the
sound or hearing is caused by the tremulation in the
cranium and its membranes ; a vibration in any one of
the bones of the cranium immediately produces a sound
within, which is also perceived as sound by the external
ear, and this on account of the close communication
that exists between the tympanum and the pericranium.
If, again, the vibration enters immediately by means of
the teeth, especially those of the lower jaw, it will similarly flow into the cranium and there produce the perception of sound ; for the ramifications of the auditory

nerve extend not only to the organ of hearing, but also
to the periosteum which surrounds the roots of the
teeth; one branch, which is soft, goes to the cochlea
and makes the membrane of the vestibulum, whence
arise a number of other membranes in the semicircular
canals, etc. ; the other branch, which is hard, runs to
the tympanum, as also to the tongue and the teeth;
branches from the first and the second cervical nerve
run also to the mouth and to the teeth, so that the
tremulation flows to and fro as on minute bridges,
finally arriving at the cartilages and membranes in the
petrous bone, and hence round about the whole cranium
and all its systems.

§ 5. If we closely examine wherein the tremulation
of sounds differs from other tremulations, we will find
that the former flows with greater swiftness and force
over all the bones and membranes than do any other
tremulations, just as when one bends a string firmly,
drawing it into a curve for a broad tremulation, whence
the sound becomes stronger than if the vibration is
moving more closely around its centre ; if we strike
anything that is expanded in the air, we will produce a
stronger sound than if we strike what lies close to
something solid.   To effect a sound in the organ of
hearing, there is needed, therefore, the following mechanism : *a*, the dura mater is expanded directly over the
cerebrum and exposed for the reception of tremulation,
whence any impulse on the dura mater immediately effects a tremulation; *b*, the membranes are in various
places affixed to the cranium, the better to arrange for
the concert and to give sufficient initiative to the sound ;
c, there are also a number of cavities or vacuities which

permit the little membranes to play with their tremulation without the hindrance of anything that is soft ; moreover, there are, in the cranium, an infinite number of porosities of such a kind as are required by the tremulation in any solid substance. The auditory part of our living organism consists, therefore, of stronger tremulations than are found in the other parts, as shall be shown further in what follows ; the difference consisting not only in the swiftness of the motion, but also in the degree of the tremulation.

* * * * * * * *
* * * * * * * * *
* * * * * * * *

CHAPTER XIII.

IT will now be necessary to make a summary conclusion of all that has been shown above, and to state
briefly wherein sensation essentially consists, as otherwise one argument may prevent the correct comprehension of the other. The case is therefore as follows : —

The first motion or tremulation takes place in the
fluid which is distributed over all the membranes, and
which flows through all the lymphatic ducts, thus crea-
ting the inmost contiguity within the body ; the least
impulse on this fluid effects a tremulation which flows
over the whole of this contiguity, with all its membranes, meninges, etc. The secondary or corresponding
cotremulation is therefore the one which takes place in
the membranes, and the third is the one that takes
place in the bony system which is so closely connected
with the membranes. From all this it follows, first,
that if the fluid is absent, there can be no impelling
force for the motion or for its communication to the
membranous system, as may be seen in cases of apoplexy or swooning ; secondly, the sanguineous fluid, if
too abundant, causes the lymphatic canals to be too
greatly expanded — they become dammed up in certain
places, or are pressed to and fro, whence the tremulatory motions are obstructed and forced to seek various
side paths, making extraordinary vibrations in the place
of the proper ones, as may be seen in cases of sudden
passion, or in drunkards ; thirdly, an insufficiency of the

sanguineous fluid causes the lymphatic canals to lie slack and flat in many places, making uneven openings for the lymph, which is thus deprived of its proper sensation, as may be seen in cases of exhaustion, fear, etc.; fourth, if the membranes are too slack, and hence wrinkled or twisted, it follows that the tremulatory fluid cannot flow forward in a proper manner, but the tremulation becomes dull or has altogether ceased before it has arrived at its terminus; fifth, if the bones are not in their proper condition, it follows that the membranes cannot be well expanded, or the lymphatic canals properly extended and open, but certain membranes must here and there fail to assist the course of the tremulation.

The summary of the whole is therefore this, that when the membranes lie well expanded towards their bones, and when the blood-vessels possess an even tension, then can everything that is fluid pass freely through all its vessels and have a contiguity throughout all the systems, so that the least impression in any place makes an impression upon all that is fluid. A vibration is thus communicated to the little valves in the vessels; a corresponding tremulation is hence imparted to the membranes, and through the latter, finally, to the bones And thus, as soon as the tremulation has been communicated to everything that is expanded and contiguous in the whole body, there results from all this what is termed a sensation. *Quod erat demonstrandum.*

It is, indeed, wonderful what a power there is in a fluid to communicate a corresponding cotremulation to a hard substance. A tremulation in the air imparts a similar motion to the whole organ of hearing, with all its cartilages and bones.   A small bomb exploded under water, causes a tremulation in the ground and the rocks round about, just as the fluids in the body communicate their tremulation to all the membranes and bones, in order to produce all that is moving and living in a man.

"As above, so below"
Expression in the macrocosm
    finds response in the
        microcosm.
From the Wisdom and Under-
    standing of the Lord comes the
    fulness of the Earth and
        the Peace and everlasting
            Power and Eternity
                in Heaven.
Thus; man and woman made
    in the image and likeness
        of our Lord God Almighty
            may by right action in
                Loving the Lord - God
                    with all his heart
                        mind, and strength
                            and his neigh
                                as himself
                                    through l
        in the Earth
            attain eternity in Heave
                There is no death, fo
                    those who seek
                        first the Kingd
                            of God and
                                righteousn
                                        ___

                            24 June 1981

INDEX

E. S — (1688–1772) · 84 years span.

SELECTIVE LIST OF SWEDENBORG'S WORKS

Philosophical and Scientific — [16 Titles and 22 volumes]

- **The Animal Kingdom,** 2 volumes
- **The Cerebrum,** 3 volumes in 2
- **Economy of the Animal Kingdom,** 2 volumes
- **The Fibre and Diseases of the Fibre**
- **Five Senses**
- **Generation. The Breasts. The Periosteum**
- **Letters and Memorials of Em. Swedenborg,** 2 volumes
- **Miscellaneous Observations**
- **Ontology, or the Significance of Philosophical Terms**
- **A Philosopher's Note Book**
- **Principia,** 2 volumes
- **Principles of Chemistry**
- **Psychologica**
- **Psychological Transactions**
- **Rational Psychology**
- **Tremulation**

Theological — [9 titles and 27 volumes]

- **Apocalypse Explained,** 6 volumes
- **Apocalypse Revealed,** 2 volumes
- **Arcana Coelestia,** 12 volumes
- **Conjugial Love**
- **Divine Love and Wisdom**
- **Divine Providence**
- **Four Doctrines**
- **Heaven and Hell**
- **True Christian Religion,** 2 volumes

Request free catalogue. Order from your book dealer or:

Swedenborg Foundation, Inc.
139 East 23rd Street
New York, New York 10010

Swedenborg Society, Inc.
20 Bloomsbury Way
London WC1A 2TH, England

General Church Book Center
Box 278
Bryn Athyn, Pennsylvania 19009